73

7٩

# MAKE it WORK!

# BUILDING

**Andrew Haslam**

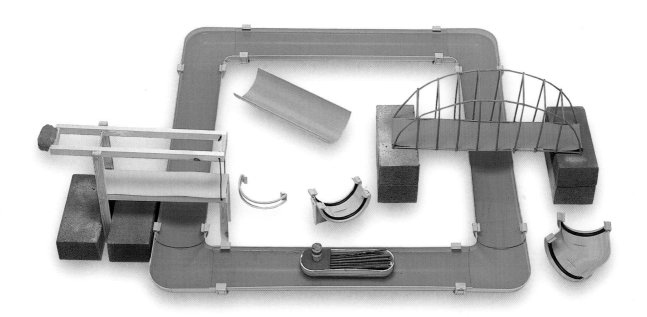

*written by*
David Glover B.Sc., Ph.D.

*Photography by* Jon Barnes

A **TWO-CAN** BOOK
published by
**THOMSON LEARNING**
New York

# MAKE it WORK!
## Other titles

Body
Insects
Machines

First published in the United States in 1994 by
Thomson Learning
115 Fifth Avenue
New York, NY 10003

First published in Great Britain in 1994 by
Two-Can Publishing Ltd.

Printed and bound in Hong Kong

**Library of Congress Cataloging-in-Publication Data**

Haslam, Andrew
  Building/Andrew Haslam; written by David Glover;
photography, Jon Barnes.
   p. cm. – (Make it work!)
  "A Two-Can book."
  Includes bibliographical references and index.
  ISBN: 1-56847-259-5
  1. Structural engineering – Juvenile literature. 2. Structural
engineering – Experiments – Juvenile literature. [1. Structural
engineering – Experiments.  2. Experiments.] I. Glover, David, 1953-
II. Barnes, Jon, ill. III. Title. IV. Series; Baker, Wendy. Make-it-work!
TA634.H37  1994
624.1–dc20                                              94-1478

Editor: Kate Asser
Series concept and original design: Andrew Haslam and Wendy Baker
Assistant model-maker: Sarah Davies
Additional design: Lisa Nutt

Thanks also to:
Colin and Jenny at Plough Studios, Rachel, Katharine and
Jonathan Bee, Ajitha Ranasinghe, Wen-Hshin Chen, Tony Ellis
and Hugo Russell-Fisher.

# Contents

Words that appear in **bold** in the text
are explained in the glossary.

Bridges, towers, domes, dams, canals –
these wonderful things are designed
and built by engineers. The engineer's
job is not easy: these structures must
carry huge loads and last for many
years, as well as look good.

## MAKE it WORK!

In this book you will learn how different
structures are put together and why they are
strong. You don't need to build a steel bridge to
see why bridge **girders** should form triangles,
because plastic straws work in just the same
way. You will discover how a canal lock works,
why an arched bridge is stronger than a flat one,
and how to turn a flat sheet into a hollow dome.

To make a strong structure, whether it
is a suspension bridge or a simple garden
wall, you must understand the **forces**
that will make it stay up. The key to
success is in understanding how to
choose the right **materials** and how
to put them together correctly.

## You will need

You can build most of the projects found in
this book out of simple materials, such as
cardboard and wood, plastic straws, and other
odds and ends. However, you will need some
tools to cut, shape, and join all the different
materials. All of the equipment above will be
very useful as part of your engineer's tool kit.

## Planning and measuring
Always plan your projects carefully before you start to build. Measure each part accurately and mark it with a pencil before you begin to cut. Mark the positions of holes before you start drilling them. A plastic measuring cup will be helpful for the projects involving water.

## Safety!
Sharp tools are dangerous. Always be careful when you use them, and ask an adult to help you. Make sure that anything that you are cutting or drilling is held firmly so that it does not slip. If you can, borrow a small table vise. It will make drilling and sawing much safer.

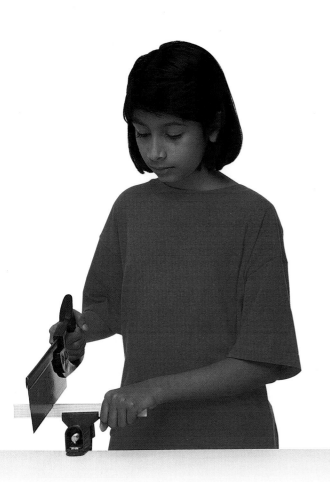

## Cutting
You will need a saw for cutting wood and scissors for cutting cardboard and paper. A craft knife is useful too, but be extra careful with the sharp blade. Always cut away from your fingers. Use sandpaper or a file to round off any sharp edges.

## Joining
There are many ways to join the parts of your structures together: a glue stick used carefully with a glue gun is one of the easiest. Different kinds of fasteners, such as nails, nuts, bolts, and Velcro, are described on page 6.

## Drilling
For some of the projects in this book, you will have to drill holes. Use a pointed awl to start the holes and then finish them off with a hand drill. This stops the drill from sliding.

The scientist Isaac Newton built a wooden bridge without using any screws, nails, or glue. After his death the bridge was taken apart for repairs, but no one could figure out how to put it back together! It was finally rebuilt with nuts and bolts.

## Push joints

Push joints are a simple way of joining materials. You can make a push joint between two plastic straws by putting glue on the end of one straw and pushing it into the end of another. Use pipe cleaners to join three or four straws, as shown in **a** below, but be careful of sharp ends. Plastic rods can be held tight with a push joint if you slide them into piping that has had a hole drilled through it, as in **b**.

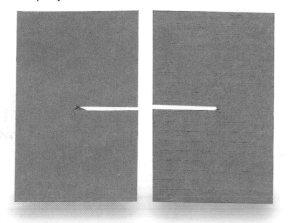

## MAKE it WORK!

Newton's bridge slotted together like a jigsaw puzzle. Usually builders need glue, screws, cement, or nails to make strong joints. Here are some ways you can make joints for your own projects.

## Staple joints

Use a hammer to tap a metal staple into a piece of wood to hold a rod or piece of cord in place. A staple gun is an excellent way of attaching cardboard to wood, but you should ask an adult to help you use it.

## Slot joints

The picture on the left shows how you can make slot joints. Cut slots in two pieces of cardboard and slide them over each other so that the pieces of cardboard cross at an angle.

## Glue joints

Glue from a glue gun is one of the quickest ways to join many items. However, be careful – the glue is very hot and you can easily burn your fingers. To make strong joints, the surfaces being glued together must be clean and dry. Use a piece of sandpaper to roughen them so that the glue can get a better grip.

## Gussets

Gussets are small pieces of a material, often metal, used to strengthen a joint. The gusset is attached across the joint. It increases the surface of the joint so that the force on it is spread out over a larger area. You can see below how you could use a cardboard gusset to strengthen a joint between two plastic straws.

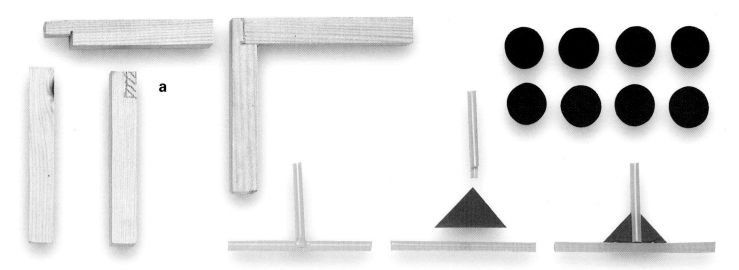

a

## Joining wood

Wood can be joined with nails, screws, or glue. If you use glue, you can make the joint much stronger by cutting slots in the wood, like the ones in **a** above. The glue then covers more of the surface of the wood than is possible when the pieces are stuck straight together.

## Velcro

Patches of Velcro behave like bristly seeds that stick to your clothes. Each bristle on the seeds ends in a tiny hook, which becomes caught in the fabric of your clothes. Velcro patches have hooks just like these. You can buy patches of self-adhesive Velcro, or ones that you stick on with glue. Velcro will make joints that you can take apart and put back together again.

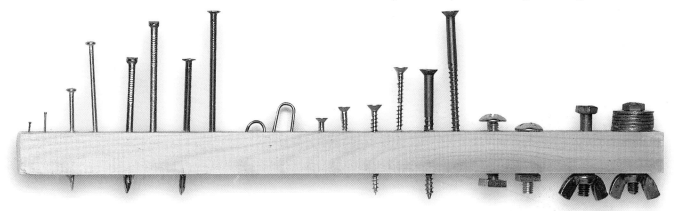

## Nuts, bolts, nails, and screws

These metal fasteners are an essential part of any builder's tool kit.

*A structure is no stronger than the joints between its parts. If the joints are weak, the structure will soon fall apart.*

It is possible to build a strong wall from stone or bricks without using cement or mortar. The wall is held together by the weight of the blocks or bricks pressing down on one another. Some very famous structures were built like this, including Stonehenge in England, the Egyptian pyramids, and Aztec temples in Mexico.

## MAKE it WORK!

The key to making a strong brick wall is the pattern in which the bricks are arranged. Build a number of walls with wooden bricks, using different patterns for the bricks. Which wall is the strongest?

**1** First make your bricks out of the balusters. They should be about $4\frac{1}{2}$ in. long. Don't worry if your strip of wood isn't exactly the suggested dimensions – just make the bricks three times as long as they are wide. Measure and mark the wood with a ruler and pencil, then cut it carefully with a saw.

**2** Sandpaper the rough edges off your bricks.

**3** You are now ready to start building. First build a wall by piling the bricks one on top of another, with the gaps in one layer directly above the gaps in the layer beneath (see below, far left). This produces a weak wall that is easily demolished.

**4** Now stack the bricks as shown in the second wall from the left. This is much more effective. With this pattern, the gaps in neighboring layers of your wall do not line up. Each brick is holding the pair of bricks below it in place and is in turn held in place by the two bricks above.

**5** Double-skinned walls, like the two below on the right, are much stronger than walls built with a single thickness of bricks. The two sides of a double wall are linked by bricks turned through right angles, known as **headers**.

## You will need

a pencil  
a ruler  
a small saw  
sandpaper  
several narrow stair balusters, about $\frac{5}{8}$" x $1\frac{1}{2}$", available at builder's supply stores.

The space or cavity in the middle of a wall of this type helps to keep a building warm, putting a protective air pocket between the inside and the outside. Sometimes the cavity is filled with foam for extra **insulation**.

# Trembling Tower

▲ A brick tower or chimney relies on the weight and pattern of its bricks to give it strength in much the same way as a brick wall. By experimenting with this tower, you will discover where the forces that keep the tower from toppling over are acting.

## How high can you build?

**1** Build a tower like the one shown above.

**2** Take bricks one at a time from the sides of the tower and place them on the top.

**3** How high can you build the tower before it collapses?

You will find that the bricks at the bottom of the tower are much more difficult to slide out than bricks higher up. This is because the weight of the bricks near the top of the tower **compresses** them. If you remove bricks from the bottom of the tower, the bricks above no longer have a firm base on which to rest, and so the tower falls over.

*The weight that presses down on the bricks at the base of a tall tower can be tremendous. The leaning tower at Pisa in Italy is falling over because its **foundations** are not deep enough to support its weight.*

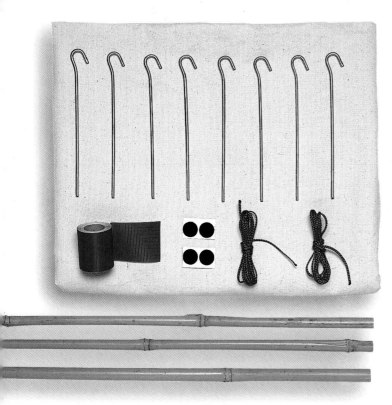

When you pitch a tent, it is kept in shape by the **tension** in the canvas and **guy ropes**. If you have a tent, you can set up house almost anywhere. Then, when you want to move on, you just release the tension, fold up the canvas, and pack it into your bag.

## MAKE it WORK!

When you stretch out a canvas tent and peg it down, tension is produced in the fabric. The upward push from the poles resists this tension. The balance of the forces pushing up and pulling down on the canvas gives the tent its rigid shape.

## You will need

| | |
|---|---|
| friends to help you | scissors |
| sticky-backed Velcro patches | tent stakes |
| strong packing or mailing tape | nylon cord |
| a large cotton or canvas sheet | |
| bamboo poles (one longer and two shorter than the sheet) | |

**1** The best place to pitch your tent is outside on some grass. Spread out the canvas sheet, making sure you have plenty of space around to stretch out the nylon guy ropes.

**2** On the outside of the sheet, reinforce the corners of the canvas with patches of tape. Turn the canvas over and place the long bamboo pole along the middle of the sheet to make the ridge of the tent.

**3** Tape the long pole in place. Using tape patches, reinforce both sides of the canvas along each edge, at the halfway point.

**4** Tie pieces of cord securely to the tops of the two shorter bamboo poles. Stick pieces of Velcro on top of each short pole and on the underside of the ridge pole at each end.

**5** Ask your friends to attach the tent poles with the Velcro, and hold the tent up while you stake the canvas down.

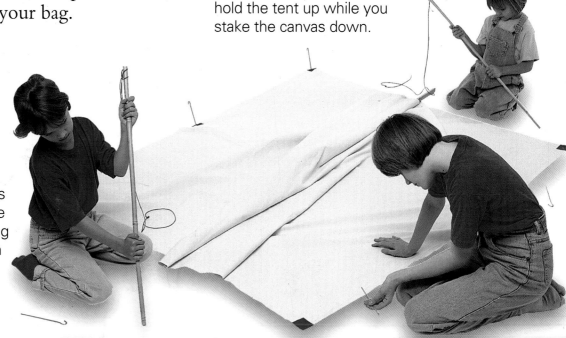

**6** Carefully stretch out the corners of the tent until it is taut, making sure that the lower edges still touch the ground. Push stakes into the grass through the four corner patches and then along the sides.

**7** Finally, stretch out the cords from the tops of the short poles to make guy ropes, and stake them firmly in place.

*Modern tents designed for expeditions are made with super-lightweight materials. A whole tent, including poles and stakes, may weigh only $4\frac{1}{2}$ lbs., yet it can withstand a blizzard and keep the people inside warm and dry.*

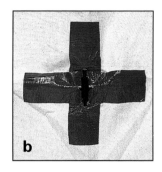

**a** **b**

Most tents need to be staked to the ground to stretch them into shape, but a hooped tent stands up without any stakes or guy ropes. Springy poles bent into hoops provide the tension needed to stretch the fabric.

## You will need

scissors                nylon cord
friends to help you     safety goggles
strong packing tape
six short, springy plastic poles
three long, springy plastic poles
a large square of canvas or cotton fabric
nine pieces of plastic tubing drilled with holes
   large enough to slide the poles through

**1** Spread the canvas on the floor and stick patches of tape on it at the 13 points shown.

**2** Using an awl, carefully make holes through the taped canvas 1¼ in. apart as shown in **a**, above.

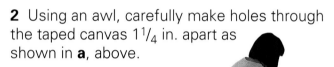

## MAKE it WORK!
The poles used for this tent are made from a very flexible plastic. You can buy them at a gardening supply store. Before modern plastics were invented, hooped tents were not common. Poles made from bamboo or metal will not bend far enough without breaking.

Make three holes in each corner patch of tape. Make four holes in the patches in the middle of each side and in the center of the canvas. In each of the remaining patches, make two holes parallel to the canvas edge.

**3** Place the long poles across the top, center, and bottom of the canvas. Position the six short poles in between to make four squares as shown.

**5** Push pieces of tubing onto both ends of the long poles, as shown in **b**, at left. Now thread the short poles through the canvas and plug them into the open ends of the tubing.

**6** Turn the tent over and fasten each pole to the canvas with tape.

**7** Tie pieces of cord to the ends of the three poles sticking out from one edge of the fabric. Push the pole-ends into the ground.

**8** Wearing the goggles, bend the poles into hoops and push the free ends into the ground. Fasten the poles firmly in place with the cord. To make your tent extra secure, you could also use tent stakes to hold the canvas down.

**Note**
The poles and cords of the finished tent may need adjusting so that the tension in the canvas is even.

**4** Thread the long poles through the holes on one edge of the canvas to the holes halfway across. Push each pole through the drilled holes in a piece of tubing, then continue to thread the pole to the other side of the canvas.

# 14 Dams

Engineers build dams to stop the flow of water in a river. The water trapped by the dam forms a reservoir that can supply water to a city. The dam can also generate electricity for the city. Water from the reservoir flows through tunnels inside the dam, which contain turbines. The turbines are turned by the water flow and drive generators to create electricity.

## MAKE it WORK!

A dam must be strong enough to withstand the huge **pressure** of the water behind it. The deeper the water, the stronger the dam needs to be. What is the strongest shape for a dam? A flat wall, or a curve? Try some experiments to discover the answer.

**You will need**

| | |
|---|---|
| colored tape | scissors |
| a large fish tank | modeling clay |
| a measuring cup | thin cardboard |

**1** Cut out a rectangle of cardboard just a little wider than your tank.

**2** Cut thin strips of tape and stick them at equal distances above each other up one corner of the tank. This scale will measure the water level in the tank.

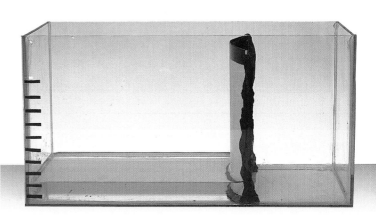

**3** To make the dam, attach the cardboard rectangle inside the tank by placing modeling clay along its sides and lower edge. Because the cardboard is slightly wider than the tank, it will curve when you fit it in place.

**4** Make sure that the clay seal is watertight all the way around the cardboard, or your dam will not work properly.

**5** Now use the measuring cup to pour water into the tank, making sure that the dam curves toward the section that contains the water.

**6** Keep filling the tank steadily, until the dam begins to leak. Make a note of the water level at this point, then continue pouring the water until the dam bursts.

How did your dam fail? Did the seal around the edges start leaking, or did the cardboard buckle and give way under the water pressure?

*People are not the only creatures to build dams. Beavers fell trees with their powerful teeth to block a stream. The lake formed behind the dam makes an excellent fishing ground for the beaver family.*

**7** Now make a dam using a rectangle of cardboard that is one and a half times the width of the tank. When you put the cardboard in place, it will curve much more than the first dam. Which dam holds more water?

**8** Finally, make another dam with a length of cardboard two or three times the width of the tank. This will make a beautifully curved dam. How much water will this dam hold? It will probably be the strongest of the three.

Can you build a dam using other materials?

*The Grand Coulee Dam on the Columbia River in Washington is the biggest concrete dam in the world. It is 1 mile long.*

A dam wall gets its strength from its curved shape. The curve can withstand forces pressing on it that would burst a flat dam wall. The curve acts in the same way as the blocks of the **arched** bridge on page 26. The more water that presses onto the dam, the tighter the blocks of the dam are squashed together.

# 16 Lock Gates

Rivers flow downhill from their source to the sea. Where the ground slopes steeply, the water tumbles in rapids and falls. A canal lock slows the flow and holds the water between the lock gates, making it possible for a boat to pass safely up or down the water levels.

## MAKE it WORK!
This model lock works like a modern canal lock. The gates slide up and down like a guillotine. In a real lock, the gates are opened and shut by electric motors.

In England there are very old locks with hinged gates that are opened and closed by hand.

## You will need
a toy boat      gravel
modeling clay      four corks
a measuring cup
pieces of plywood
a long, narrow water tank
four grooved wooden strips, bought from a craft supply store (or eight short strips glued together in pairs)

**1** Stand the wooden strips upright inside the tank and attach them with the modeling clay as shown. If necessary, seal the edges of the lock with clay to make them watertight.

**2** Cover the bottom of the tank with gravel. Make the gravel layer thicker on one side of the lock than on the other to make different depths of water.

**3** Ask an adult to help you cut the lock gates from the plywood. The gates should slide smoothly up and down in the grooves of the strips. Drill a finger hole at the top of each gate and two holes for corks at the bottom.

**4** Fit corks into the lower holes and slide the gates shut.

**5** Now pour some water into the tank. Half fill the inside of the lock and the end with least gravel. Make the water level at the other end of the tank much deeper than this.

## Operating the lock

To operate the lock, first float a boat at the end of the tank with the lower water level, so that the boat can pass through into the lock. When the boat is inside, lower the lock gate. Now you must raise the water level in the lock so that the boat can float out to the upper level.

Remove one cork from the second gate, or lift the gate slightly. Add more water to the end section of your tank to simulate the water coming from upstream.

When the water in the lock is level with the water beyond it, lift the second gate for the boat to pass through. Can you figure out how a boat passes down hill through a lock?

Cable cars carry passengers high up into the mountains, where there are no roads or trains. The **gondolas**, or cars, hang from a huge loop of thick, steel rope strung between two towers. To take the weight, both the towers must be extremely strong. They also need to be set in firm foundations so that they do not topple over.

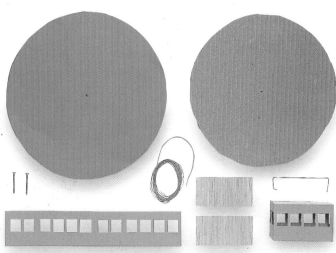

## MAKE it WORK!

Use your tower-making skills to build this model cable car. The gondolas run between towers made from strips of balsa wood. The strips are stuck together in triangles to give the towers the strength to carry the load.

## You will need

| | |
|---|---|
| two screws | string |
| strips of balsa wood | thin wire |
| thick corrugated cardboard | colored tape |
| thin cardboard and strong glue | modeling clay |

**1** Cut two rectangles of balsa wood to make the roof and floor of each gondola.

**2** Make the gondola walls from thin cardboard. Cut holes for the windows as shown.

**3** Assemble the gondolas. Glue the walls to the edges of the roof and floor. Tape a small piece of bent wire to the roof so that you can hang the gondola from the cable.

**4** Now build the towers. Arrange the balsa-wood strips in triangle shapes as shown. Make one tower two stories high, and the other four stories. Secure the joints with strong glue.

**5** Glue small cardboard gussets around the joints to strengthen them (see page 7).

**6** Cut squares of balsa wood to fit the tops of each tower and glue them in place. Strengthen them with a gusset at each corner.

**7** To make the large wheels around which the cable turns, draw two circles on corrugated cardboard and cut them out. Then cut four slightly larger circles out of thin cardboard. Glue each small, thick circle between two large, thin ones.

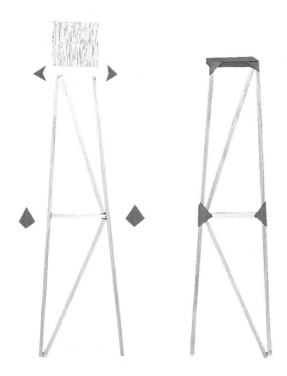

*The cables on a real cable car are kept slack so that there is as little tension in the cable as possible. If a cable were stretched too much, it would pull on the towers or run the risk of snapping in a strong gust of wind.*

▼ Turn one of the cable wheels to make the gondolas travel up and down between the towers.

**8** Make a hole in the center of each cable wheel. Push screws through the wheels and twist them into the balsa-wood squares on top of the towers. Make sure that the cable wheels turn freely on the screws.

**9** To keep the towers from toppling over, you must attach them to the floor or weight them down at the base with modeling clay. Real towers are set in massive concrete foundations.

**10** Loop the string around the wheels and tie the ends together. Do not pull the string too tight, or the tension will damage the towers.

**11** Hang the gondolas from the cable. Use a small piece of tape to keep them from sliding down the slope.

*The world's longest cable car is built in two sections. It carries passengers nearly 4 miles into the Swiss Alps.*

Some shapes are naturally stronger than others. For instance, a square is very weak. A square frame, made from four pieces of wood nailed together, can easily be squashed into the shape of a diamond. Triangles are much stronger than this. Three pieces of wood joined together in a triangle are rigid and will not twist out of shape. When bridges and towers are built from tubes or girders, the parts are put together in triangles to give the structure strength.

**You will need**

| | |
|---|---|
| an awl | a board |
| a vise and a drill | a tape measure |

two pieces of strong nylon rope, about 5½ ft. and 4½ ft. long

three pieces of wooden broom handle, each about 3 ft. long

**1** Ask an adult to help you drill a hole about 1 in. from each end of each pole.

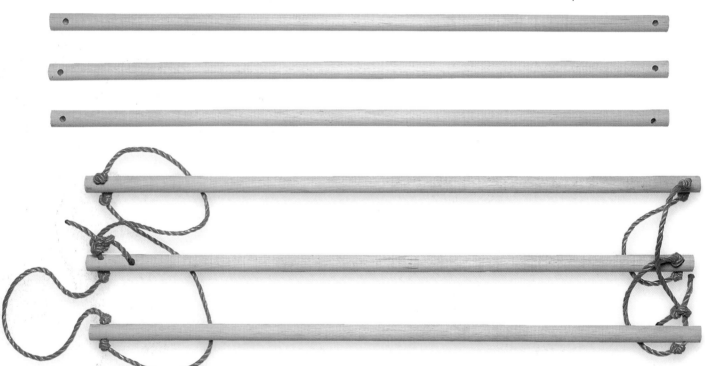

## MAKE it WORK!

This stool is made from just three wooden poles and two pieces of rope. There are no nails or screws, yet it can hold the weight of a person. The stool gets its strength from its shape. The poles and ropes are arranged in triangles. These triangles fit together to make two rigid pyramids, one on top of the other.

**2** Now fasten the poles together with the ropes as shown. First tie a thick knot 4 in. from the end of one rope. Thread the long end of this rope through a hole in the first pole.

**3** Tie a second knot in the rope on the other side of the hole, so that the pole is held firmly between the two knots.

**4** Secure the second pole between two knots in the same way, leaving approximately 16-20 in. of rope between the poles.

**5** Leave free an identical length of rope and tie the third pole in place.

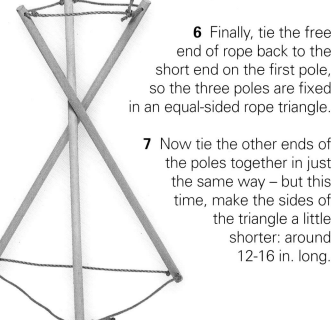

**6** Finally, tie the free end of rope back to the short end on the first pole, so the three poles are fixed in an equal-sided rope triangle.

**7** Now tie the other ends of the poles together in just the same way – but this time, make the sides of the triangle a little shorter: around 12-16 in. long.

### ▲ Setting up the stool
Once all three poles are roped together at both ends, you are ready to set up your stool. It can be tricky, so you'll probably need to ask a friend to help you.

Pull all three poles apart so that the rope triangles are stretched out. Stand the larger triangle on the floor, and then twist the smaller triangle around until all three poles cross in one place. Now rest a board on top and take a seat!

### Tension and compression
When you sit on the stool, your weight pushes down on the poles and compresses them. The tension in the ropes keeps the poles from sliding and keeps the stool rigid.

*A bicycle frame is built in a triangular shape, so that it does not twist or buckle when you put your weight on it.*

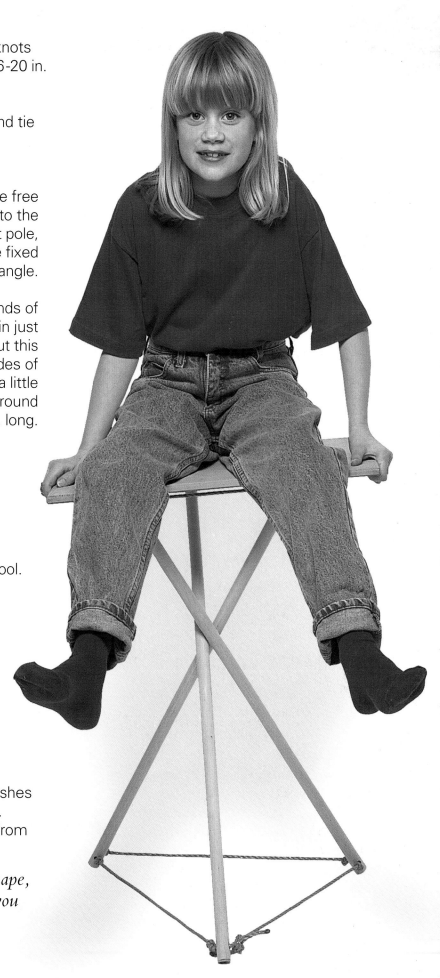

Have you seen a tower crane lifting girders high over the rooftops in the city or loading cargo onto the ships at the docks? The swinging **boom** on top of the tower moves the load carefully into position.

## You will need

three empty thread spools     string
small plywood squares          gravel
pieces of dowel
a yogurt cup
thumbtacks
paper clips
a glue gun

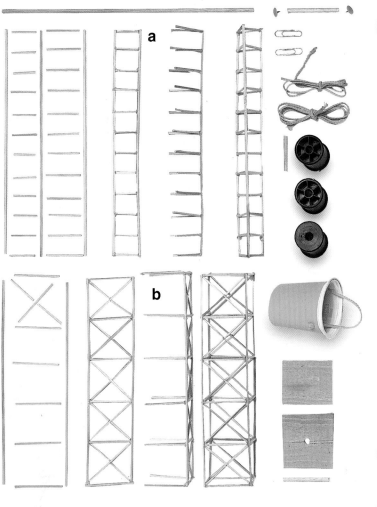

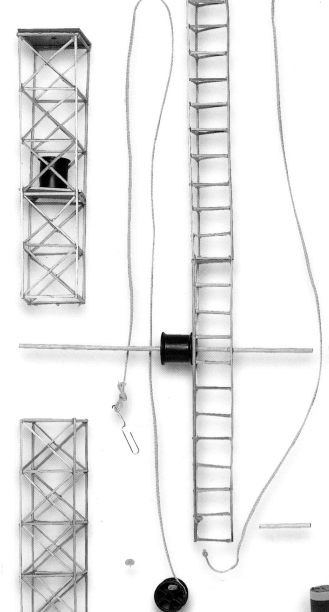

**a**

**b**

## MAKE it WORK!

A tower crane must be strong and light so that it is easy to move. The boom must also be well balanced so that the tower does not topple over when the crane lifts a load. The boom of this model crane is balanced with a yogurt cup that is filled with gravel. The crane is constructed from dowel girders arranged in triangles for strength. The girders of a real crane are made of steel.

**1** To make the three-sided boom, first glue dowel pieces $1\frac{1}{4}$ in. long between two dowel pieces $9\frac{1}{2}$ in. long, to make a flat ladder (see **a** left).

Now glue more short dowel pieces to either side of a third piece of dowel, in a "V" shape. Join this structure to the flat ladder to complete the boom.

**2** For the tower, make two box frames 2 in. square and $9\frac{3}{4}$ in. tall from dowel pieces (see **b** left).

**3** Cut a plywood square with 2-in. sides and another square a little smaller. In the middle of the 2-in. square, drill a hole large enough for a dowel rod to pass through.

**4** Glue a thread spool to the center of each square, matching the holes. Stick the 2-in. square on top of one box frame, then stick the smaller square inside the same frame (see right).

**5** For the **winch**, glue a short dowel rod to the end of the boom as shown. Put the thread spool on the dowel and push a thumbtack into the dowel end, to hold the spool in place.

**6** Join the boom and tower with a piece of dowel as shown, then hang the yogurt cup from the boom as a **counterbalance**.

**7** Thread and fasten the strings as shown. Use a dowel to make a handle for the thread spool winch to lift the load.

You will need to adjust the position of the counter-balance according to the size of the load. If the crane topples forward as it lifts the load, move the counter-balance back. In a real crane, the counterbalance adjusts itself automatically as the load is moved by the boom.

*A large tower crane can lift a load of about 30 tons to a height of 500 ft.*

Have you ever had to wait in traffic for a drawbridge to close? A drawbridge moves to allow tall ships to pass along a river or a canal. Some bridges, called swing bridges, swing sideways to open. Others, like the one shown here, move up and down.

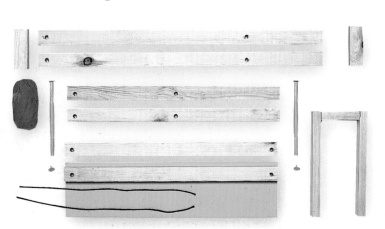

**MAKE it WORK!**
This simple drawbridge operates using a counterbalance. The weight of the **deck** is balanced by the weight on the **lever arm**.

**You will need**
pieces of dowel
strips of wood
stiff cardboard
modeling clay
thumbtacks
a glue gun
string
a drill

**1** Cut the wood to length as shown. You will need two pieces about 8 in. long, four pieces 6 in. long and two pieces 2 in. long. Drill holes at the positions indicated.

**2** To make the deck, cut a piece of cardboard measuring 2 in. x 6 in. Glue two of the 6-in. strips of wood along its edges.

**3** Cut a piece of dowel into two rods about 3½ in. long. Make two 2-in. cardboard sleeves for the rods. Line up the holes in one end of the deck with those midway along the two remaining 6-in. strips. Slide a dowel rod through the holes and push thumbtacks into the ends of the rod to keep it in place.

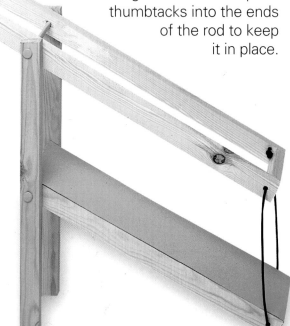

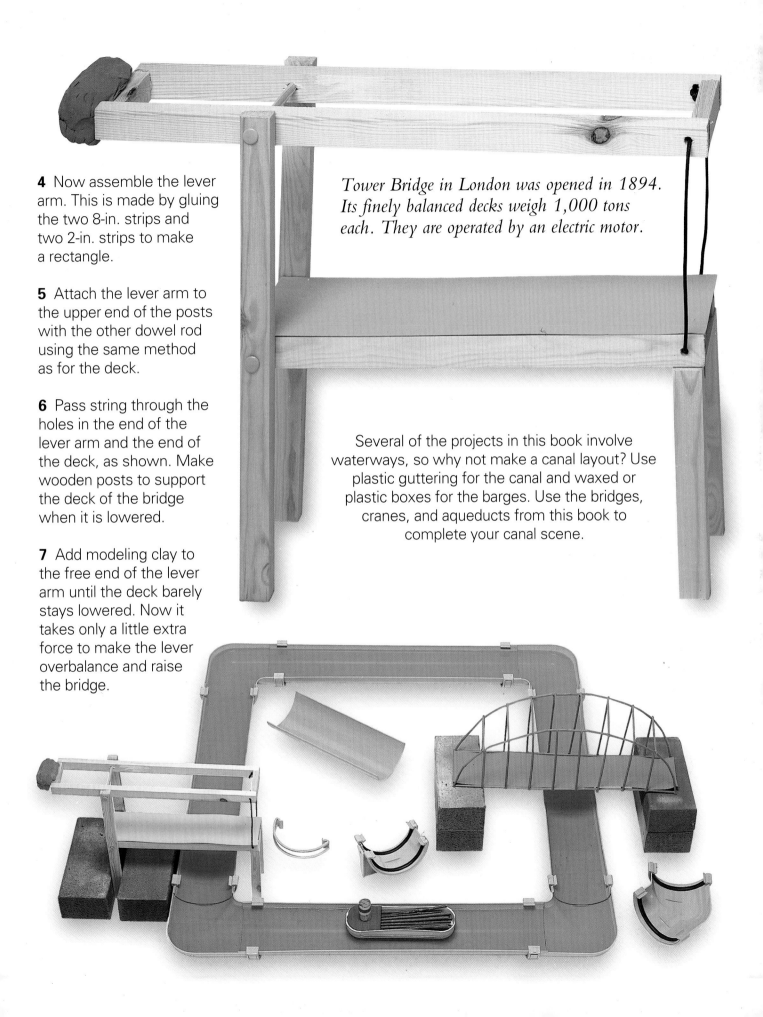

**4** Now assemble the lever arm. This is made by gluing the two 8-in. strips and two 2-in. strips to make a rectangle.

**5** Attach the lever arm to the upper end of the posts with the other dowel rod using the same method as for the deck.

**6** Pass string through the holes in the end of the lever arm and the end of the deck, as shown. Make wooden posts to support the deck of the bridge when it is lowered.

**7** Add modeling clay to the free end of the lever arm until the deck barely stays lowered. Now it takes only a little extra force to make the lever overbalance and raise the bridge.

*Tower Bridge in London was opened in 1894. Its finely balanced decks weigh 1,000 tons each. They are operated by an electric motor.*

Several of the projects in this book involve waterways, so why not make a canal layout? Use plastic guttering for the canal and waxed or plastic boxes for the barges. Use the bridges, cranes, and aqueducts from this book to complete your canal scene.

How can a bridge made from blocks or bricks hold together without glue or cement? A flat brick bridge would fall apart under its own weight alone. However, by arranging the blocks in an arch you can produce a structure of great strength that holds itself in place.

### MAKE it WORK!
To build an arch you will need blocks that have sloping ends that fit together to make a curve. Wooden crates would be best to use, but you could make your own blocks from cardboard.

### You will need
crates with sloping sides
two friends to
    help you
bricks

**1** Try to judge how long your bridge will be. Place two piles of bricks this distance apart on the floor to keep the ends of the arch from sliding apart. Move them in or out if the sides of the arch will not meet.

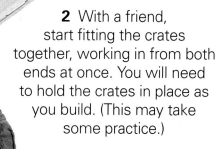

**2** With a friend, start fitting the crates together, working in from both ends at once. You will need to hold the crates in place as you build. (This may take some practice.)

### Note
Make sure the crates you haven't used yet are within easy reach!

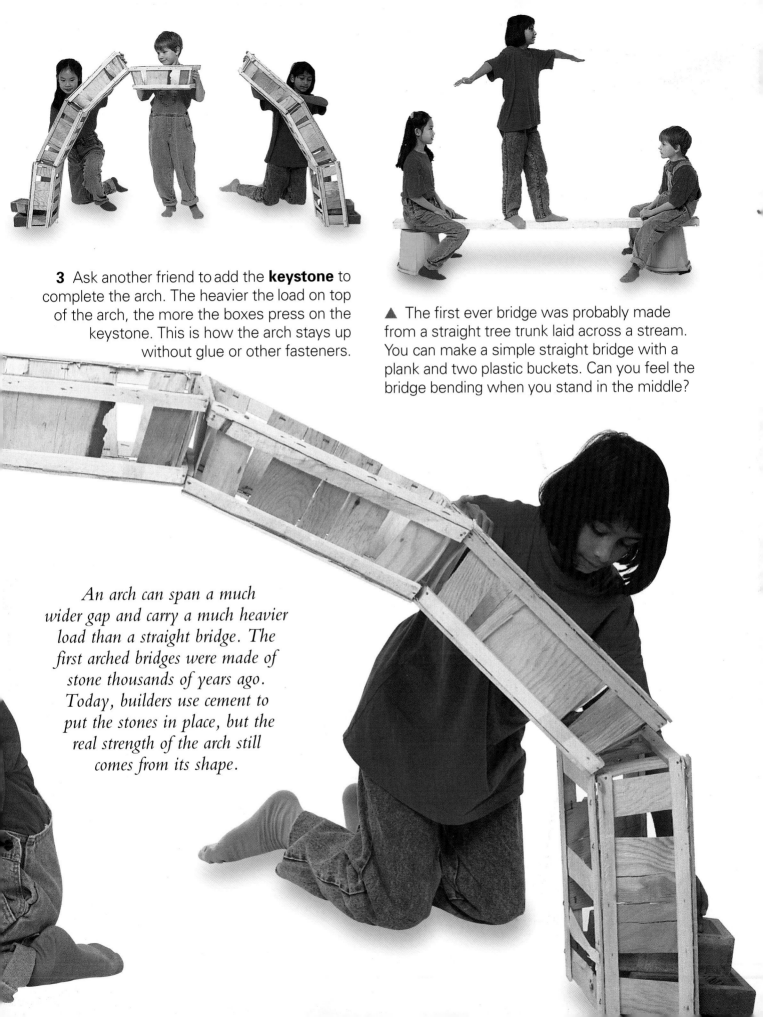

**3** Ask another friend to add the **keystone** to complete the arch. The heavier the load on top of the arch, the more the boxes press on the keystone. This is how the arch stays up without glue or other fasteners.

▲ The first ever bridge was probably made from a straight tree trunk laid across a stream. You can make a simple straight bridge with a plank and two plastic buckets. Can you feel the bridge bending when you stand in the middle?

*An arch can span a much wider gap and carry a much heavier load than a straight bridge. The first arched bridges were made of stone thousands of years ago. Today, builders use cement to put the stones in place, but the real strength of the arch still comes from its shape.*

Cars and trains on bridges are familiar sights, but a boat crossing a bridge is more of a surprise. Yet this is just what happens with an aqueduct. The strong arches that support an aqueduct carry a canal full of water above the ground.

## You will need

plastic guttering with end caps
a sharp knife    stiff cardboard
a metal ruler    a baseboard
bricks    glue

**1** The arches of this aqueduct are each made from a strip of stiff cardboard that measures about 4 in. x 13 3/4 in. Cut the strips to size with the ruler and sharp knife. Be careful with the knife.

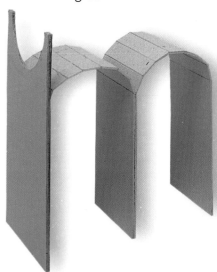

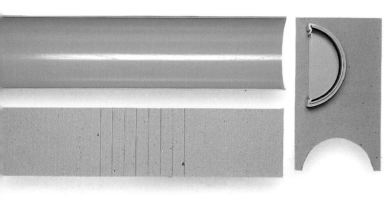

### MAKE it WORK!

Use a row of cardboard arches to support the weight of a plastic canal in your own aqueduct.

**2 Score** eight lines 3/4 in. apart across the middle of a strip of cardboard. Bend the strip to make a shallow arch with two supports.

**3** Take another strip of cardboard, and on it draw a semicircle, using the end of the gutter as a guide. Cut along the line to leave a curve in which the guttering can rest.

*The ancient Romans were highly skilled engineers. They built amazing aqueducts to supply their cities with water. Many of these aqueducts are still standing 2,000 years later.*

**4** Glue together a pair of arches and a gutter support. Your aqueduct is assembled from three or four of these basic units.

**5** Put the aqueduct together on a baseboard. If necessary, support the walls at either end with bricks.

The arches of your aqueduct support the weight of the canal evenly along its length. But what is the best design for the arches? Should they be flat or very pointed? Experiment with different shapes of arches to see which is the strongest and which is most stable.

Stone and brick are strong, but they are very heavy. For a long bridge, which has to support its own weight as well as the loads that cross it, lighter materials are needed. Engineers use steel girders, linked together in **lattice** patterns, to build long bridges that are strong, light, and not too expensive.

**1** First make the bow-shaped sides of the bridge using pieces of plastic straw. The piece that you bend to make the bridge's arch should be about one and a half times as long as the girder (the bowstring) that runs along the bottom.

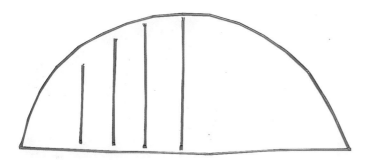

## MAKE it WORK!

Bowstring bridges are named after the string of an archer's bow. Tension in the bridge's bottom girder – the bowstring – holds the arch in place, just as tension in the string curves a bow. Find out how the **ties** carry the strength of the arch to the deck by building your own bridge.

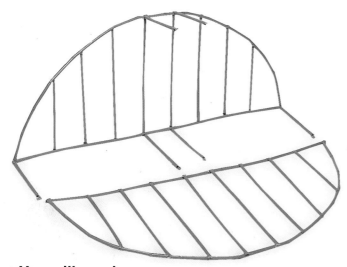

### You will need

bricks
scissors                    glue
plastic straws          thin cardboard

**2** You can join pieces of straw by putting a drop of glue on the end of one piece and pushing it into the end of another. You may have to split the end of one straw a little way.

**3** To attach the ends of the bowstring girder to the arch, split the ends of the bowstring straws. Open out the ends to make small flaps, then cut one of the flaps off each end. Glue the remaining flaps to the ends of the arch. Use this method to attach the ties between the arch and bowstring.

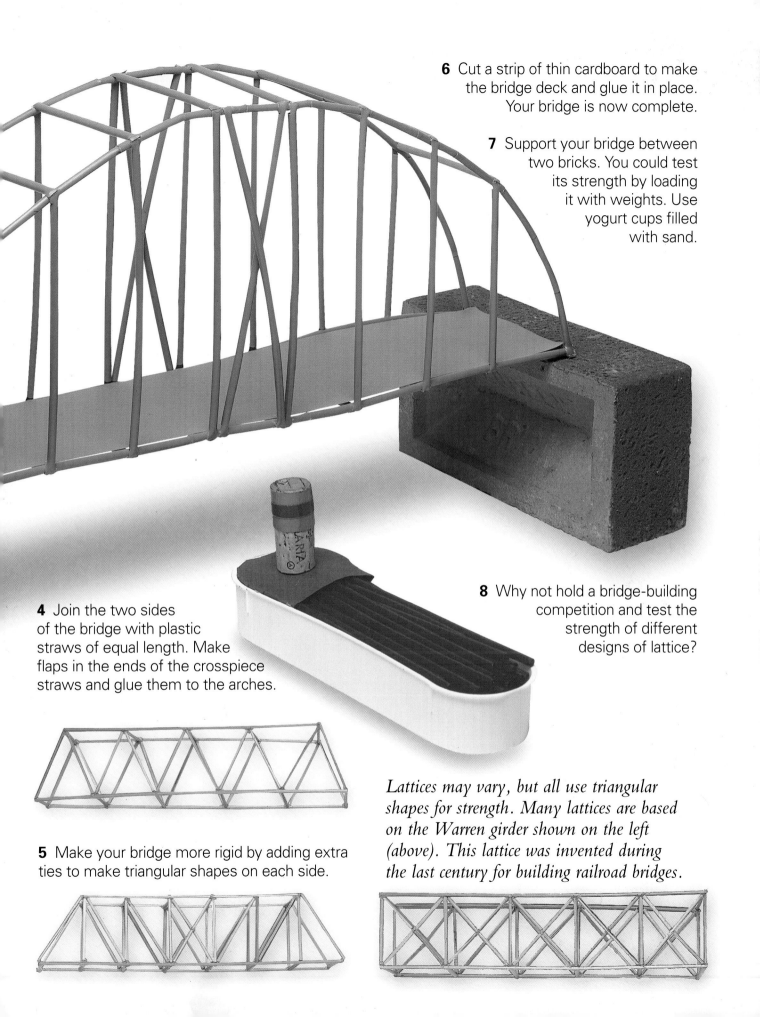

**6** Cut a strip of thin cardboard to make the bridge deck and glue it in place. Your bridge is now complete.

**7** Support your bridge between two bricks. You could test its strength by loading it with weights. Use yogurt cups filled with sand.

**8** Why not hold a bridge-building competition and test the strength of different designs of lattice?

**4** Join the two sides of the bridge with plastic straws of equal length. Make flaps in the ends of the crosspiece straws and glue them to the arches.

**5** Make your bridge more rigid by adding extra ties to make triangular shapes on each side.

*Lattices may vary, but all use triangular shapes for strength. Many lattices are based on the Warren girder shown on the left (above). This lattice was invented during the last century for building railroad bridges.*

# 32 Suspension Bridges

A good way to cross a river is on a rope tied between two trees. This simple idea is used to build the world's longest, most elegant bridges. The weight of the traffic crossing a suspension bridge is carried by the two massive cables strung between the towers of the bridge.

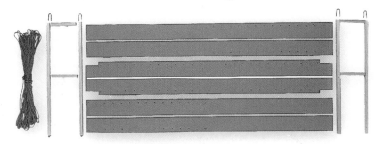

## MAKE it WORK!

The cables in a real suspension bridge must carry thousands of tons of weight. They are wound from strong steel wires and can be over 3 ft. thick.

The cables in this model are made from nylon string, but the same principle is at work.

## You will need

| | |
|---|---|
| six bricks | strips of cardboard |
| wire staples | string |
| strips of wood | glue |

**1** Cut four 8-in. wooden strips and four 2½-in. strips. Glue them together to make the towers, as shown.

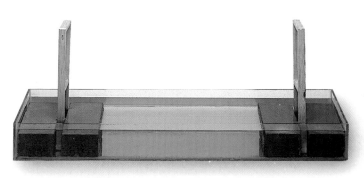

**2** Stand the towers about 16 in. apart using four bricks to keep them in place.

**3** Cut three strips of cardboard measuring 16 in. x 2½ in. to make the main deck and the two approach ramps.

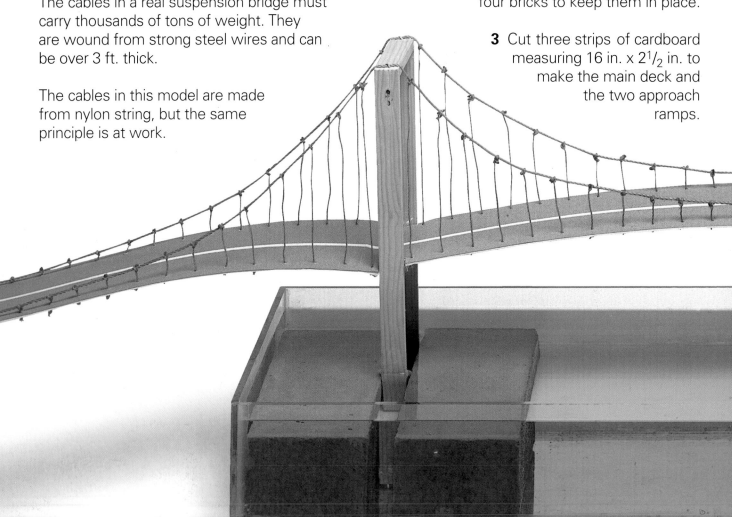

**4** Punch small holes every ³/₄ in. along both edges of the cardboard strips. Use an awl or better still, if you have one, a leather punch.

**5** For the bridge cables, stretch two long pieces of string over the towers, attaching them to the tops of the towers with staples. Place the spare bricks on either side of the bridge and tie the ends of the strings around them.

**7** Finally, tie pieces of string every ³/₄ in. along the cables of the bridge, as shown. Pass them through the holes in the deck of the bridge and knot them tightly underneath. For extra strength, adjust the lengths of string to make the deck flex into an arch.

You could use your bridge and a water tank to make a river crossing like the one below.

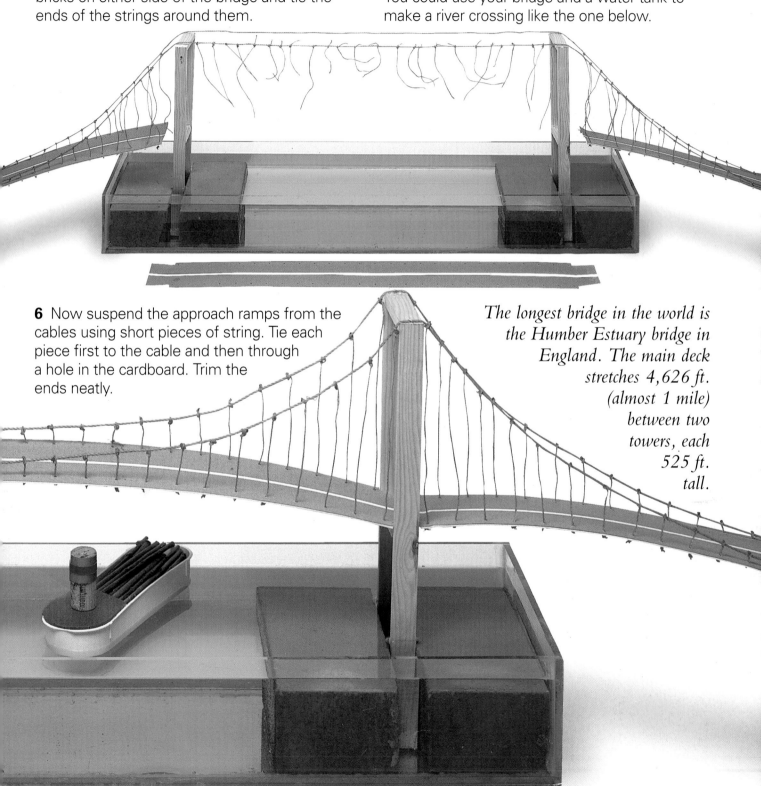

**6** Now suspend the approach ramps from the cables using short pieces of string. Tie each piece first to the cable and then through a hole in the cardboard. Trim the ends neatly.

*The longest bridge in the world is the Humber Estuary bridge in England. The main deck stretches 4,626 ft. (almost 1 mile) between two towers, each 525 ft. tall.*

Making a strong waterproof roof can be the most difficult part of building a house. Flat roofs often leak. A sloping roof works better because the water runs off, but the roof must be strong enough to support the weight of heavy tiles and to stand up to high winds.

**MAKE it WORK!**
The roof of a modern house is made from a framework of triangles called **trusses**. The sloping sides of each truss are called **rafters**. The rafters are linked across the bottom by a **tie beam**. A series of trusses can be linked together to build a roof of any length.

**You will need**
cardboard
a glue gun          a knife
thin dowels        plastic straws

**1** Cut six pieces of straw with angled ends as shown. Glue the three longer pieces together into a triangular truss. Glue the three shorter pieces inside, as shown below, to give the truss extra strength.

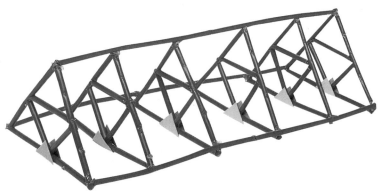

**2** Reinforce the joints at the base of the truss with small cardboard gussets.

**3** Make five more identical trusses. To join them together in a roof frame, glue the trusses to three long straws by their points. Make the trusses an equal distance apart.

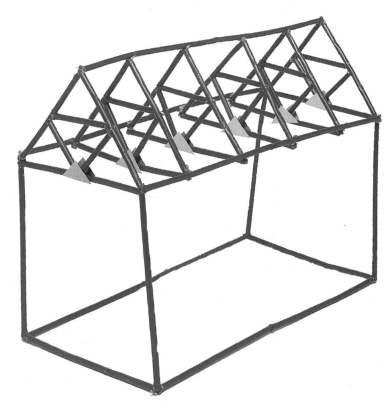

**4** Use more straws to build the framework of a house to support your roof.

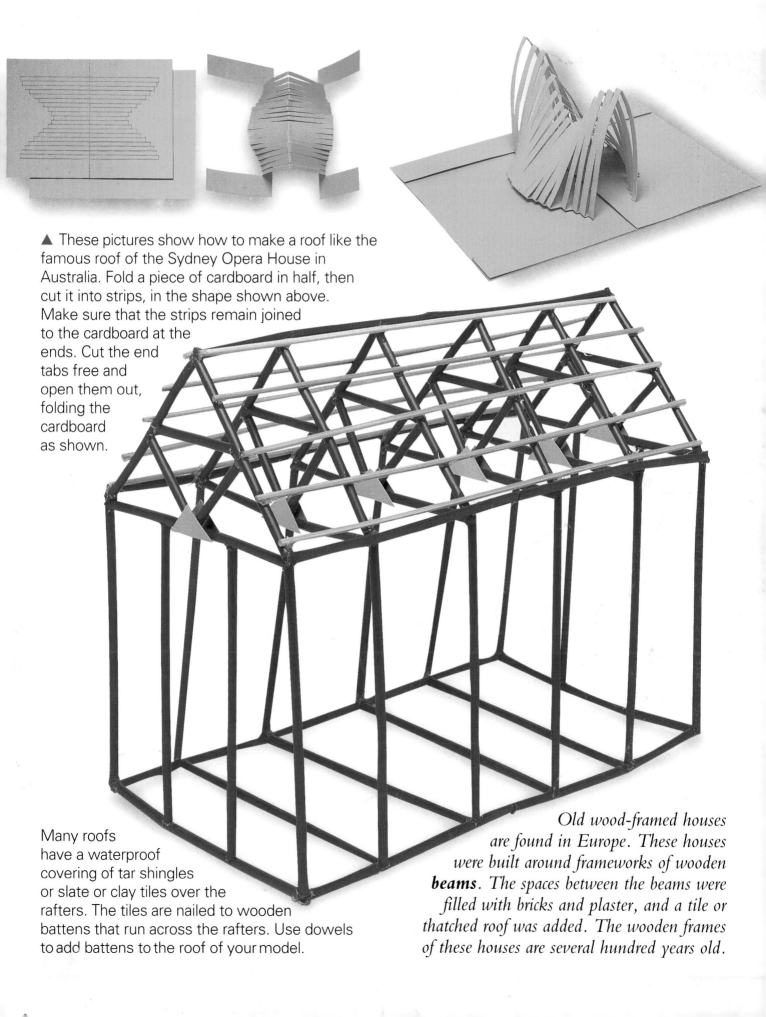

▲ These pictures show how to make a roof like the famous roof of the Sydney Opera House in Australia. Fold a piece of cardboard in half, then cut it into strips, in the shape shown above. Make sure that the strips remain joined to the cardboard at the ends. Cut the end tabs free and open them out, folding the cardboard as shown.

Many roofs have a waterproof covering of tar shingles or slate or clay tiles over the rafters. The tiles are nailed to wooden battens that run across the rafters. Use dowels to add battens to the roof of your model.

*Old wood-framed houses are found in Europe. These houses were built around frameworks of wooden* **beams**. *The spaces between the beams were filled with bricks and plaster, and a tile or thatched roof was added. The wooden frames of these houses are several hundred years old.*

Stone Age people built their homes out of the natural materials around them. Using only a flint knife, they cut sticks, tied them into a frame with dry grass, and then covered the frame with a skin to make a tent.

Some sticks will need to be tied so that they cross. The pictures above show you how to do this using a short piece of string. Some sticks must be tied lengthwise. The picture along the bottom of the page shows you how to overlap the sticks and tie them with two knots.

## MAKE it WORK!

Make your own Stone Age tent with flexible sticks tied into a dome-shaped lattice.

### You will need

long flexible sticks (plant stakes from a
  gardening supply store are ideal)
safety goggles
scissors
string

**1** Start by making a ring of sticks for the base of the tent. Tie sticks together as shown below, bending them as you go.

**2** When you have a complete ring, lay it on the ground. Wearing the goggles, take long sticks, or short ones tied together, and make hoops at intervals across the ring. Tie the ring about 4 in. from the ends of the hoops. The hoops should meet above the middle of the ring.

**3** Tie the hoops together where they meet at the top of the dome.

**4** Finally, strengthen the dome lattice by tying three or four stick rings around it at different heights. The more rings there are, the stronger your dome will be.

Throw a sheet of canvas over your dome framework to turn it into a cozy tent.

Instead of using string, you could tie the sticks with the plastic ties used by electricians to fasten cable together. You can buy these from a hardware store.

**Note**
Be careful not to poke your eyes with the stick ends.

▲ This small model was made using the same method as the lattice dome below. It shows clearly how the sticks fit together.

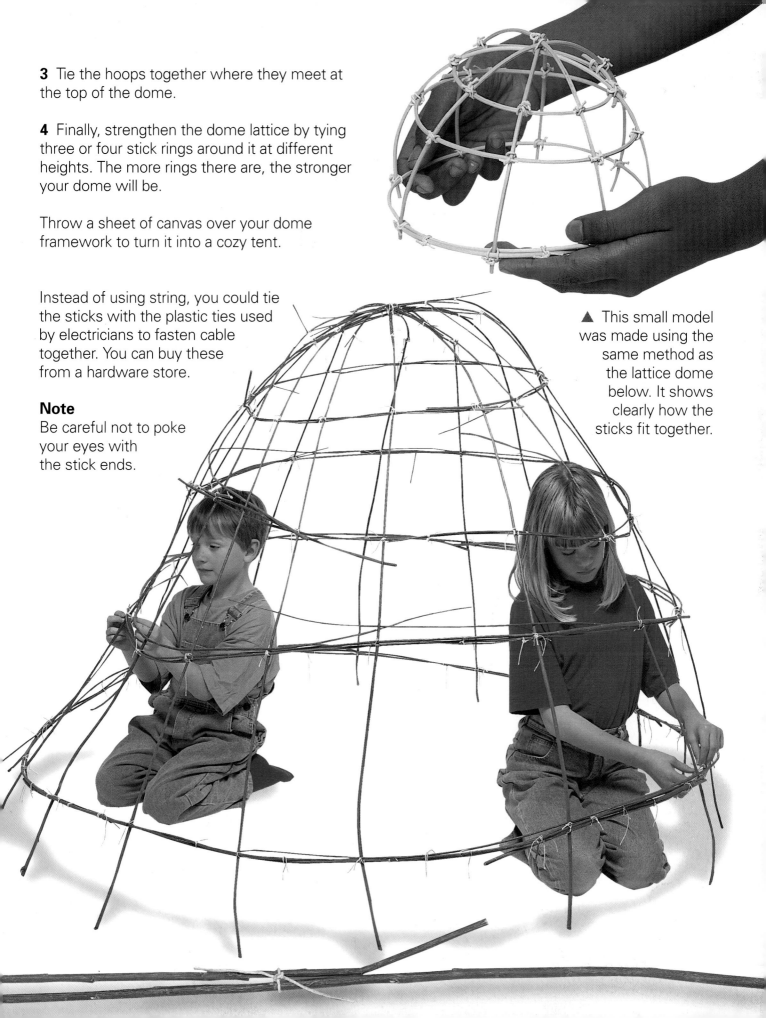

Cardboard boxes are made by folding flat sheets of cardboard and gluing the edges together. But what shape should the cardboard sheet be in order to make a cube-shaped box? Try taking a box apart to find out. The flat sheet that this box was built from is made up of six squares joined at the edges.

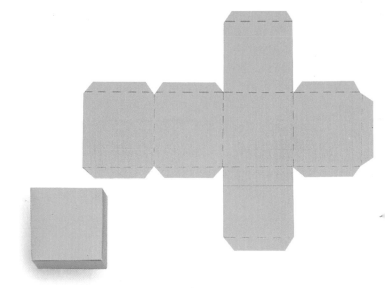

### MAKE it WORK!
The flat starting shape used to make a box is called a **net**. The solid shape it makes when it is folded is called a **form**. The examples on this page show how different nets fold up to make different solid forms. Try making some of them yourself.

### You will need
a pair of drawing compasses
a pair of scissors
thin cardboard
a pencil
a ruler
glue

**1** Start by making a cube. Draw six squares arranged as shown above. Add the nine flaps.

**2** Cut out the cube net and its glue flaps.

**3** Fold the net along the dotted lines and glue the flaps inside the edges of the squares where they meet.

When you have completed the cube, try to make some more difficult shapes.

A pyramid made from four equal-sided triangles (see far left) is called a tetrahedron. The blue pyramid with the square base is the same shape as the pyramids of Egypt.

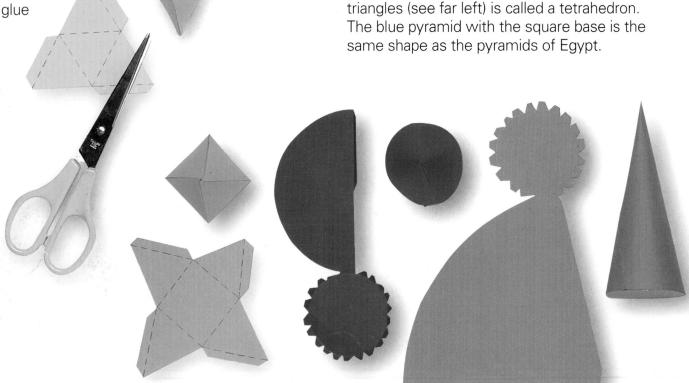

To make cones, use your compasses to draw circles joined to semi-circles, as shown. Experiment to see how changing the radius of each curve affects the cone's shape.

▼ The long nets below fold into shapes called prisms. The three-sided shape is a triangular prism and the six-sided one is a hexagonal prism.

▲ The six-pointed net above folds to make a six-sided pyramid. The four-pointed star folds into a tall shape like an Egyptian **obelisk**.

Make a mobile out of some of your shapes by hanging them from pieces of thread attached to dowels. You could even make your own gift boxes.

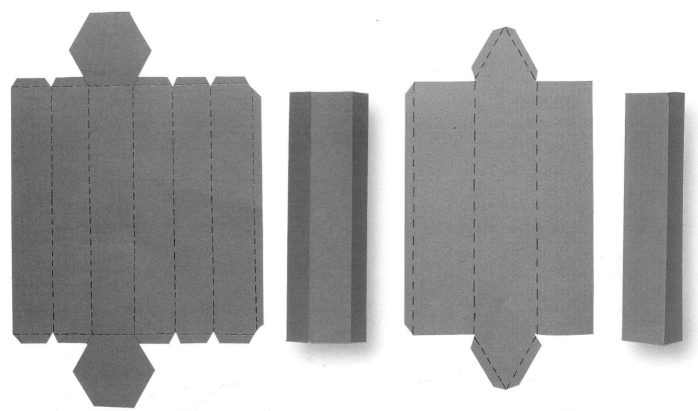

Try peeling an orange, keeping the skin in one piece. Flatten out the peel onto a board. What does it look like? The skin tears into several parts with curved edges, which are joined together. The shapes below are like the flattened-out orange peel. They are called polyhedrons, from the Greek word meaning "objects with many sides."

## MAKE it WORK!
The nets below are like the flattened orange skin. They fold up into shapes with so many sides that they are almost spheres.

### You will need
| | |
|---|---|
| cardboard | a protractor |
| a ruler | scissors |
| glue | a pencil |

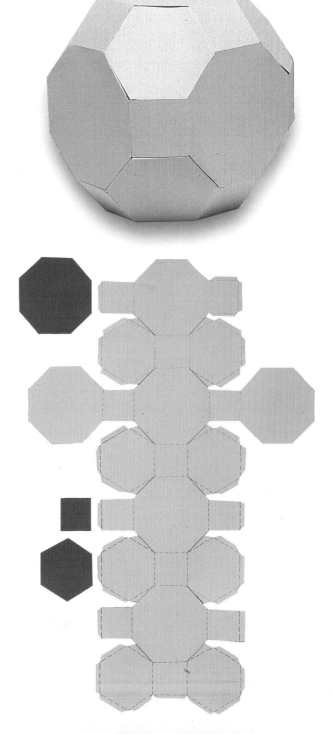

**1** Make cardboard templates for the basic shapes that make up each net, shown in red cardboard below. (You can use a protractor for this or trace the red shapes and enlarge them.)

**2** Draw around your templates to make the net for your polyhedron.

**3** Add the glue flaps to your net. Finally, fold up the net along the dotted lines and glue the flaps in place. This may take some patience!

*The third polyhedron from the left is made from pentagons and hexagons. This pattern is often used to make soccer balls.*

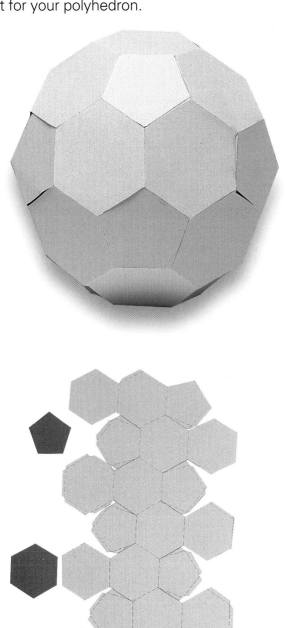

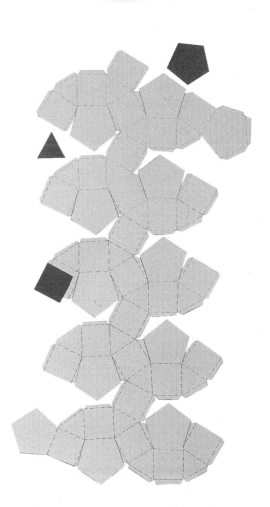

An Inuit's igloo is made from blocks of ice fitted together to make a dome. Domes are strong and hold heat well because their surface is small compared with the space inside. Fifty years ago, the designer Buckminster Fuller saw that large domes can be made by joining together simple flat shapes.

## You will need

| | |
|---|---|
| an awl | a screwdriver |
| a pencil | a sharp knife |
| wing nuts and screw | a steel ruler |
| large sheets of strong cardboard | |
| an adult and friends to help you | |

**1** With an adult, measure and cut out the shapes below. The sides of each shape should measure 2 ft. You will need ten triangles with flaps on each side, and six pentagons. Put one pentagon aside to be the roof.

**2** Cut a large hole in one of the remaining pentagons to make a door.

### MAKE it WORK!
The igloo on this page is made entirely out of two simple shapes – triangles and pentagons – joined together.

**3** Join a triangle to one side of each of the five remaining pentagons. Fasten the pieces together with screws and wing nuts. Use the awl to make holes for the screws. (If you prefer you could use Velcro patches instead of screws to join the shapes together.)

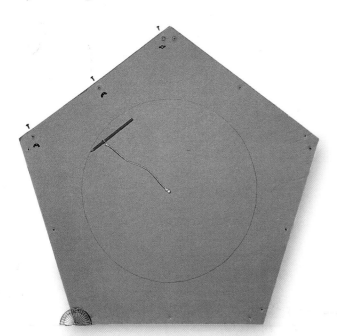

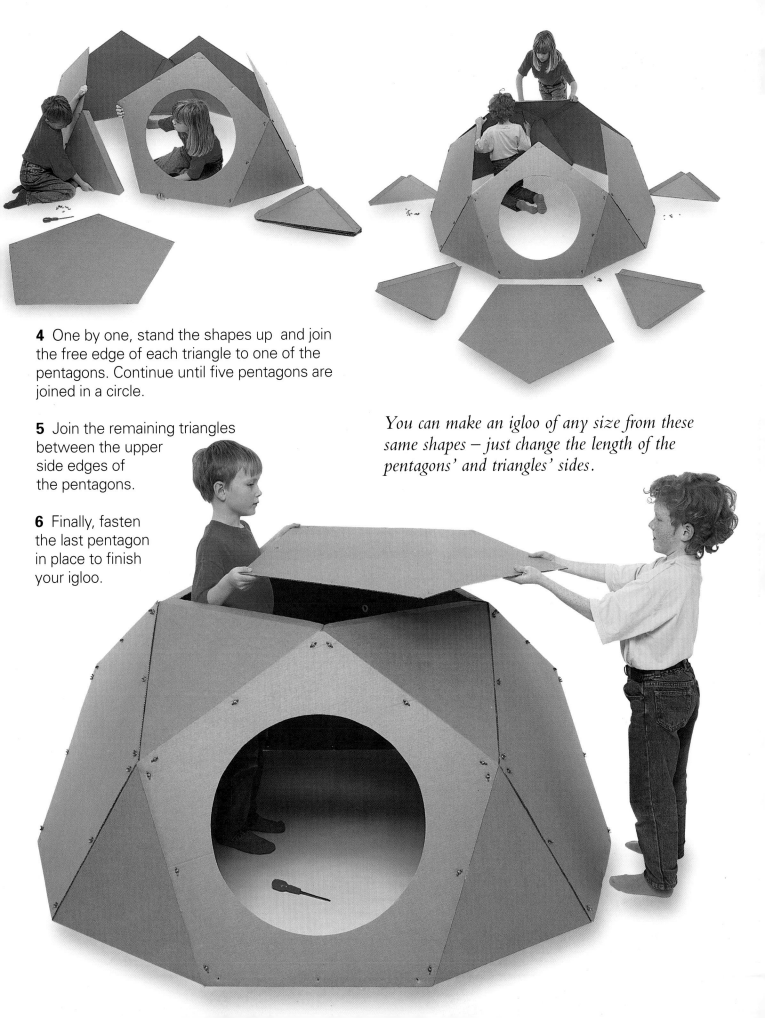

**4** One by one, stand the shapes up and join the free edge of each triangle to one of the pentagons. Continue until five pentagons are joined in a circle.

**5** Join the remaining triangles between the upper side edges of the pentagons.

**6** Finally, fasten the last pentagon in place to finish your igloo.

*You can make an igloo of any size from these same shapes – just change the length of the pentagons' and triangles' sides.*

Building in outer space is difficult. Astronauts have to wear clumsy spacesuits, and all the things they use must be transported from Earth by rocket. It takes a lot of fuel to carry heavy materials into space. **Modules** made up of simple shapes weigh less and are easy to build.

**You will need**

a sharp knife  a pencil
thick cardboard  a ruler
a friend to help
screws and wing nuts or Velcro patches

**1** Cut out and assemble the parts of an igloo as described on pages 42-43.

**2** Build a second igloo.

### MAKE it WORK!

The igloo on page 42 is made from just two simple shapes. By joining two igloos together, you can make yourself a space module. You could then link several modules together with short tunnels to build yourself a space station. The different modules could be used for eating, sleeping, and working.

**3** Turn one of the igloos upside down.

**4** Lift the second igloo and place it on top of the first. Line up the pentagons of the top half with the triangles of the bottom half.

**5** Fasten the two igloos together with wing nuts and screws.

Your space module is now complete.

*NASA (National Aeronautics and Space Administration) is planning to build a space station using modules. The lightweight shapes that are joined together to build the space station will be carried into space by the space shuttle* Discovery.

**Arched** Built in a curve. When a force such as the weight of traffic or water presses on an arch, the blocks of the arch are pressed together and held more firmly in place.

**Beams** Long pieces of wood or metal used by engineers and builders to support a load.

**Boom** A long pole or girder used to position loads, which can swing on a hinge at one end of the pole. A boom is used for a yacht's sail as well as the load on a crane.

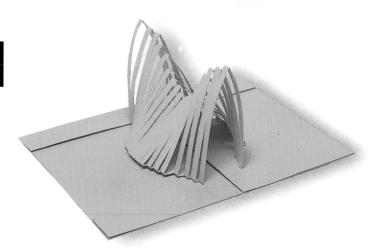

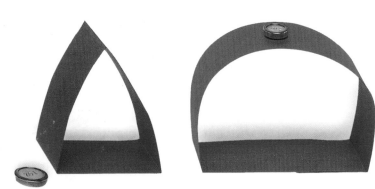

**Compresses** Squeezes together. The force of compression squashes a cushion when you sit on it. The bricks at the base of a tall tower are compressed by the weight of the bricks above them pushing down.

**Counterbalance** A weight used to balance the load being lifted by a lever arm.

**Deck** The floor or platform of a bridge, over which traffic passes.

**Forces** Pushes or pulls that try to move an object. Forces on a building either squeeze, stretch, or twist its parts. If the structure of the building is not strong enough, the parts may snap or buckle.

**Form** A shape, for example one created by folding up a paper net.

**Foundations** The parts of a building beneath the ground. Foundations spread the weight of the building over a larger area and reduce the pressure on its base.

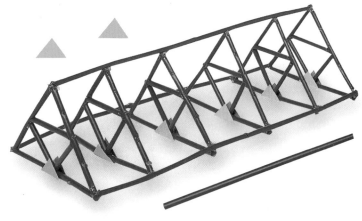

**Girders** Beams made from iron or steel.

**Gondolas** The cabins in which the passengers travel on a cable car.

**Guy ropes** Ropes used to steady the poles and canvas of a tent.

**Headers** Bricks laid at right angles to the face of a wall.

**Insulation** Foam or other materials that keep heat from escaping from a building.

**Keystone** The topmost stone of an arch, which holds the rest in place.

**Lattice** A structure made by joining rods or beams into regular patterns. Lattices made from triangles are usually chosen for building, because triangular shapes are strong.

**Lever arm** A hinged arm that is used to lift heavy loads.

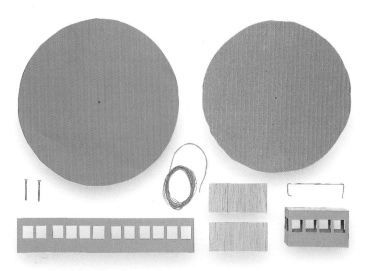

**Materials** Substances, such as plastics, woods, and metals, that are used to make other objects.

**Modules** Simple parts, often made up of regular shapes, that are joined together to build a complicated structure.

**Net** A flat sheet, made up of simple shapes joined together, that can be folded up into a solid shape or form.

**Obelisk** A tall, pointed block of stone. In Egypt, obelisks were carved with prayers and names.

**Pressure** The pushing force that acts on an object when it is surrounded by air or water. You can feel the pressure on your ears when you swim underwater. The deeper the water, the greater the pressure.

**Rafters** The sloping beams that form the frame of a roof.

**Score** To make cuts or lines in an object, so that it will bend more easily. Score lines can be made using the blade of a pair of scissors or a knife and a metal ruler.

**Tension** A stretching force.

**Ties** Strings or rods that prevent two parts of a structure from separating.

**Tie beam** A beam that holds together the ends of two rafters in a roof frame.

**Trusses** Frameworks used to support a roof frame. Trusses are often arranged in triangles.

**Winch** A machine that lifts a load by winding up a rope or chain attached to the load.